THE GREENHOUSE EFFECT

BY
JACK C. HARRIS

CRESTWOOD HOUSE
New York

Collier Macmillan Canada
Toronto

Maxwell Macmillan International Publishing Group
New York Oxford Singapore Sydney

Library of Congress Cataloging-in-Publication Data
Harris, Jack C.
 The greenhouse effect / by Jack C. Harris. — 1st ed.
  p. cm. — (Earth alert)
  Includes bibliographical references
  Summary: Discusses the causes of the greenhouse effect, what it does to our planet, and how it can be stopped.
  1. Greenhouse effect, Atmospheric—Juvenile literature. [1. Greenhouse effect, Atmospheric. 2. Global warming.] I. Title. II. Series.
QC912.3.H37   1990   363.73'87—dc20   90-36294   CIP
ISBN 0-89686-543-6   AC

Photo Credits
Cover: Earth Scenes: (Doug Wechsler)
Earth Scenes: (Zig Leszczynski) 4; (Anthony Bannister) 9; (Dr. Nigel Smith) 15
Devaney Stock Photos: 7, 13, 33; (Bruce Berg) 10, 39; (P. Grant) 20; (Don L. Black) 28;
 (George J. Hardy) 31
Animals Animals: (Doug Allan) 17; (C.C. Lockwood) 25
AP—Wide World Photos: 22
Superstock: 36; (A. Azzarello) 27; (Daniel T. Staples) 40

Copyright © 1990 Crestwood House, Macmillan Publishing Company

All rights reserved. No part of this book may be reproduced or transmitted in any form or by any means, electronic or mechanical, including photocopying, recording, or by any information storage and retrieval system, without permission in writing from the Publisher.

Macmillan Publishing Company   Collier Macmillan Canada, Inc.
866 Third Avenue   1200 Eglinton Avenue East
New York, NY 10022   Suite 200
                                  Don Mills, Ontario M3C 3N1

Produced by Flying Fish Studio Incorporated

Printed in the United States of America

10 9 8 7 6 5 4 3 2

CONTENTS

Introduction...5
What Are the Greenhouse Gases?.......................................11
Causes of the Greenhouse Effect..12
Warning Signs...16
Is the Greenhouse Effect Real?...18
A Strange, Warm World..23
The Greenhouse Effect and Daily Life...............................26
Stopping the Greenhouse Effect...30
Making New Laws..37
What You Can Do...41
For More Information...43
For Further Reading..44
Glossary..45
Index...47

Just as glass walls trap heat in a greenhouse, gases in the atmosphere can hold too much of the sun's heat close to the earth.

INTRODUCTION

Imagine that you fall asleep for a long time. You don't wake up until the year 2010. When you wake up, you look at a newspaper. There you see strange headlines:

NEW YORK: GREENHOUSE EFFECT CAUSES FLOODING

IOWA: GREENHOUSE EFFECT DESTROYS CORN CROP

FLORIDA: GREENHOUSE EFFECT PUTS HOUSE UNDER WATER

You would probably have some questions about the greenhouse effect. What is it? Where did it come from? Why didn't you hear about it back in the 1990s? Is it as bad as the headlines make it sound?

To understand what the greenhouse effect is, you must first know about greenhouses. A greenhouse is a heated building for growing plants. Greenhouses protect plants from wind and weather. A glass roof and walls help keep it warm inside, even in winter. Another name for a greenhouse is a hothouse. The "greenhouse effect" means that the earth might become hot like a greenhouse.

The earth's surface has land and oceans, rivers and plains and mountains. Above the surface is a blanket of gases. This blanket of gases is called the atmosphere. The air we breathe is part of the atmosphere.

The earth is heated by rays from the sun. Much of the sun's heat is absorbed by the earth's atmosphere. Some is reflected back into space. Many scientists fear that changes in the atmosphere will make the gases in it act like the glass roof and walls of a greenhouse. The atmosphere will trap more heat near the earth. This would cause the earth to heat up too much. That's why they call it the "greenhouse effect." The greenhouse effect is sometimes known as global warming.

The main cause of the greenhouse effect is carbon dioxide. Carbon dioxide is a gas that absorbs heat from the sun. You can't see or smell it. But it is a natural part of the atmosphere. It is abbreviated and written as CO_2.

Carbon dioxide is in the air we breathe every day. In every ten thousand parts of air, there are only three or four parts of carbon dioxide. We may not even notice it.

Carbon dioxide is formed in many ways. One common way is by burning. Coal, wood, and oil all contain the element carbon. So do many kinds of food, such as sugar and bread. When any of these are burned, the gas carbon dioxide is formed. It escapes from the burning substance into the air.

In nature, carbon dioxide is formed by fermentation. Fermentation is the slow decay of sugar or things that contain sugar. Grapes are fermented to make wine, for example.

Green plants need carbon dioxide to live. Plants take water from the soil and carbon dioxide from the air. Sunlight gives them energy. The plants use the energy to combine the carbon dioxide and water. This makes a kind of sugar that plants need to live.

Smoke produced by factories contains large amounts of carbon dioxide, a main cause of the greenhouse effect.

After making the sugar, the plants have oxygen left over. They release this oxygen into the air. This process is called photosynthesis.

People need oxygen to breathe. We breathe in oxygen and breathe out carbon dioxide. Plants need carbon dioxide to make their food. They take in carbon dioxide and release oxygen. This is part of the balance of nature that makes life on earth possible. Both carbon dioxide and oxygen are essential to life on earth.

So carbon dioxide is not a bad gas. The problem, scientists say, is in the balance. Carbon dioxide absorbs lots of the heat rays from the sun. If there is too much carbon dioxide in the air, the earth will heat up. The planet will become too hot. It will be like a greenhouse.

The greenhouse effect is a scientific theory. A theory is an explanation of how something works. When more information is available, a theory may prove to be true. For example, hundreds of years ago some people thought the earth was flat. They believed in a flat earth theory. Other people thought the earth was round. They believed in a round earth theory. Now we know that the earth is round. The round earth theory is true. The flat earth theory is false.

Some theories have not yet been proved true or false. The greenhouse effect theory is one of these. It describes some things that are happening now. It says that these events will cause other things to happen in the future.

Many scientists believe the greenhouse effect is a good theory. They think it accurately explains what is happening. Some other scientists disagree. They argue that no one knows yet if more carbon dioxide in the atmosphere will make the earth too hot. In this book, you can read what scientists on both sides say about the greenhouse effect. You can decide what you believe.

During photosynthesis, plants take carbon dioxide from the air and change it into oxygen, which is then released back into the atmosphere.

Methane, a greenhouse gas, is produced by animals like cows during the digestive process.

WHAT ARE THE GREENHOUSE GASES?

Scientists call carbon dioxide a greenhouse gas. That means that it absorbs infrared heat rays from the sun. The sun has more than one kind of ray. Infrared rays are heat rays that may cause the greenhouse effect. Ultraviolet rays are another kind of sun ray. They are the rays that cause sunburn.

Carbon dioxide is by far the most important greenhouse gas. But there are others. You may have been put to sleep at your dentist's office by laughing gas. Laughing gas is really nitrous oxide. This is another greenhouse gas. Nitrous oxide is produced by burning coal and by the exhaust from motor vehicles. Some fertilizers may also produce this gas. About five million tons of nitrous oxide go into the atmosphere each year. As far as the greenhouse effect is concerned, nitrous oxide is very dangerous. That is because nitrous oxide absorbs 250 times as much heat as carbon dioxide.

Many animals make methane, another greenhouse gas. Cows produce methane in their stomachs when they eat. When a cow burps, it exhales methane. Termite mounds produce methane. So do rice paddies and swamps.

Chlorofluorocarbons are another greenhouse gas. They are also called CFCs. CFCs are used in air conditioners, refrigerators, and freezers. They are also in some plastic foam products, spray cans, and solvents. There are very few CFCs in the atmosphere — only 200 to 400 parts per trillion. Still, they are a very dangerous greenhouse gas. CFCs absorb 10,000 times as much of the sun's heat as carbon dioxide. Unlike carbon dioxide, CFCs do not exist naturally. They are all produced by human activity.

CFCs are also dangerous to the earth's ozone layer. Ozone is a kind of oxygen. A layer of ozone exists eight to thirty miles above the earth. This ozone layer protects the earth against the ultraviolet rays of the sun.

Other ozone exists much lower in the atmosphere — two to five miles above the earth. At the lower level, ozone acts as a greenhouse gas. Some scientists believe that the lower level of ozone may have increased by 25 percent in the last hundred years. This is the time since the Industrial Revolution, when people started using machines and a lot of coal, oil, and gas to run them.

CAUSES OF THE GREENHOUSE EFFECT

There are two ways humans cause the greenhouse effect. The first way is by putting more greenhouse gases like carbon dioxide in the air and upsetting the balance of nature. We put carbon dioxide in the air when we burn coal, gas, and oil in factories. More

Carbon dioxide is added to the air whenever coal, gas, or oil is burned.

carbon dioxide goes into the air when cars burn gasoline. We also put greenhouse gases in the air when we use CFCs in refrigerators and Styrofoam cups. When people clear land by burning forests—or even burning garbage—more carbon dioxide is released.

The second way people help cause the greenhouse effect is by keeping greenhouse gases in the air. They do this chiefly by destroying forests. Forests help remove carbon dioxide from the air. They also put oxygen back. Some people call forests the "lungs" of the earth. When forests are destroyed, there are fewer trees to turn carbon dioxide into oxygen. More carbon dioxide stays in the air.

Destroying forests is called deforestation. Some forests are destroyed by fire. Others are cut down to make room for houses and factories. Or they are logged for wood.

Huge areas of North America used to be covered by forests with trees more than one thousand years old. Now, only 15 of every 100 trees are left. In addition, 27 million acres of rain forests are destroyed each year.

Today, tropical rain forests cover one-fifth of the earth. Soon, some scientists say, there will not be enough trees left to remove carbon dioxide from the air. It will stay in the atmosphere forever, adding to the greenhouse effect.

Deforestation destroys trees that are needed to remove carbon dioxide from the atmosphere.

WARNING SIGNS

When did scientists start to think about the greenhouse effect? Is this a new threat, or have they known about it for a long time? If they knew about it, why have people waited so long to do something to prevent it?

The possibility of a greenhouse effect was predicted long ago. In 1896, Svante August Arrhenius said that if the amount of carbon dioxide in the atmosphere was doubled, global warming would occur. Arrhenius was a Swedish chemist, physicist, and Nobel Prize winner. His work and theories were very close to today's thinking about the greenhouse effect. But there was one big difference. Arrhenius didn't think global warming was dangerous. To him, the warming meant that another ice age wouldn't take over the earth. Since Arrhenius's day, studies have shown that the amount of carbon dioxide has been increasing.

In 1958 scientists began some important tests. One test took place on an extinct volcano in Hawaii. Scientists chose this site, high above a frozen lava sea, because the air tests would not be affected by plants or traffic.

The first tests showed that there were 315 parts of carbon dioxide for every million parts of air. Now, more than 30 years later, there are 348 parts per million. That is an increase of 33 parts per million in 30 years.

Scientists also tested air trapped in bubbles in glaciers. This air is more than a hundred years old. The "old air" shows only 280 parts of carbon dioxide per million. By comparing the "old air" to air tested during the last 30 years, scientists show that we are now putting much more carbon dioxide in the air.

Other scientists have made computer models of the earth's atmosphere to study warming trends. A study at NASA's Goddard

By studying air trapped in glaciers, scientists have been able to measure the increase in the amount of carbon dioxide in the atmosphere over the last century.

Institute showed a very slow warming. In the last one hundred years, average earth temperature rose about one-half degree Celsius. This is about one degree Fahrenheit. (Scientists use the Celsius scale to measure temperature. Most Americans use the Fahrenheit scale.)

Rising sea levels are another indication that the greenhouse effect is changing the world. During the past hundred years, records have shown that the seas have become deeper by a small amount. About 25 percent of this increase is due to normal heat expansion. The other 75 percent is a mystery. Some scientists say this is because ice is melting at the North and South Poles. They believe this melting is caused by the greenhouse effect.

Are these really warning signs? Scientists do not all agree. Some say that these signs do not prove anything. They say these are normal climatic changes. Others are quite sure that the greenhouse effect is real.

IS THE GREENHOUSE EFFECT REAL?

If the greenhouse effect is real, then, scientists say, change is needed right away. Every year that we delay will mean a warmer world. It took years to build up this much greenhouse gas in the atmosphere. It will take years to stop putting more gas there. And it will take even longer to try to reduce it.

Many scientists urge quick action. They say that the federal government should pass laws to reduce greenhouse gases. Such laws might order car companies to build only cars that save on fuel. They might require factories to stop burning fuels that release so much carbon dioxide. They might say that only certain kinds of fuel may be burned.

These scientists say it will take many years to clean up the air, so we should start now. The longer we wait, the worse the greenhouse effect will be.

Many business people are against strict new laws, however. They say such laws would make some companies shut down. Then many people would lose their jobs. They fear that the damage to the economy would be far worse than the damage to the environment.

Electric power companies that burn a lot of coal are also concerned. Some scientists say coal-burning is a major source of greenhouse gases. So the power and coal companies are worried the government will pass laws restricting the use of coal. They send lobbyists to Washington. Lobbyists are paid to try to influence lawmakers. The power companies' lobbyists argue that power and coal companies are not causing the greenhouse effect. Some lobbyists say that there is no greenhouse effect at all.

Many scientists agree with them. They point out that the earth has always had cool times and warm times. They claim that recent warm years were just one of the naturally warm times. Other people point out that four of the hottest years in this century came in the 1980s and say this is a sign of global warming. But these scientists argue that four years is a very short time in the earth's history. Since human beings have been recording temperatures

The 1980s had some of the hottest temperatures in the 20th century. Some scientists believe this is a sign of the greenhouse effect.

for only the last few hundred years, no one really knows if the earth has been this warm before.

Some scientists also question the tools used to research the greenhouse effect. They say that "models" of the atmosphere are not accurate. They say the models leave out important facts, like the effect of the oceans and clouds. Some scientists believe the models make the threat seem worse than it really is.

Time magazine reported that James Hansen, a NASA official, told members of the United States Congress that he believed the greenhouse effect had already arrived. Since he stated that, Hansen's findings have been challenged by many other scientists. One of them was Andrew Solow, a statistician at the Woods Hole Oceanographic Institution's Marine Policy Center. He noted that the computer models used to predict the coming of the greenhouse effect could not even detect the small .5-degree Celsius rise in the earth's overall temperature that has occurred in the last 100 years.

Many scientists agree that the decade of the 1980s was the warmest in the century. But they don't agree on its cause. While environmentalists say it was due to global warming, others point to the period between 1940 and 1965. During these years, the earth experienced a cooling trend. At this time, the burning of coal, oil, and gas was higher than in the 1980s. These facts seem to dispute the fears of global warming based on increased industrial activity.

Another group of scientists say that a greenhouse effect is possible. But, they argue, it is not yet dangerous. Gradual warming of the earth will not hurt people or plants or animals for thousands of years.

Not all scientists agree that the world is growing warmer. But those who believe in the greenhouse effect think our new warm world will be a very strange place.

Global warming would produce great changes on our planet. In Iowa, for example, farmers could no longer grow corn—the soil would be too dry.

A STRANGE, WARM WORLD

Imagine prairies turning into vast deserts. Picture forests turning into tropical jungles. Think of the North Pole melting and the oceans rising. These are some of the changes that environmentalists say could happen with the greenhouse effect — perhaps by the middle of the 21st century.

Here's how some environmentalists think the world will look in 2050 with the greenhouse effect.

In Louisiana, people will worry about floods. Much of Louisiana is low land, near the ocean. As the ice caps at the North and South Poles melt, the ocean will rise. As the ocean rises, many homes will be flooded. Whole towns will have to move.

Water will be a worry in the southwestern United States, too. With more heat and less rain, people won't have enough water to drink. Many will have to move.

New York City will have hotter summers. But that is not the worst problem: The Hudson and East rivers will flood, too.

In Iowa, farmers will have to stop growing corn. The summers will be too hot and dry. Kansas will look like a desert. Farmers in Minnesota and Canada and Siberia will be able to grow more corn and wheat than before. Their summers will be longer with global warming.

Many of the forests in Minnesota and Canada will die. The trees will not be able to change to live with more heat. Many of the lakes and swamps will also dry up. These changes will leave wild animals without homes. Ducks and other birds will be confused by changes in the climate. They will not know when to lay eggs or when to fly south for the winter. Many of their nest sites will disappear.

In the Nile Valley in Egypt, many farmers will lose their land. Floods will drive them away, but they will have no place to go.

The Sahara Desert will grow larger. Hungry people along its edges will have even less food. More and more people will starve in Africa.

The rich countries in Europe and North America will suffer. People will have to drive less, use less air conditioning, learn to live with less electricity. But people in poor countries will suffer more. In these countries, people will starve.

Global warming will change life for plants and animals, too. Whole forests will disappear. Birds that live in these trees will no longer have homes. Animals that live in forests and find their food there may starve.

As the oceans rise, fish will suffer, too. Temperature changes will kill many fish. Plants that grow under water may die. Then some kinds of fish will no longer have food.

Changes in the weather can also cause animal mutations. New kinds of animals and plants might begin to appear on the earth.

These are just some possible changes. Scientists can only guess at others. But whether all these changes happen or not, almost all scientists who fear the greenhouse effect agree that the world will be a totally different place for plants, animals, and humans if we don't take some serious steps today.

If temperatures rise all over the earth, including the ocean, marine life will suffer, too.

THE GREENHOUSE EFFECT AND DAILY LIFE

Human beings are adaptable. We can make changes in our lives to adjust for changes in nature. We have already begun to make some of these changes.

Not long ago, people raked their lawns in the fall. Then they piled up the leaves and burned them. Today, burning leaves is against the law in many parts of the country. These laws protect the air from pollution by leaf fires.

Our garbage system has been changed, too. There are now laws against burning garbage. And there are laws about how to bury garbage safely. Recycling is becoming more and more common. People are learning to recycle cans and glass and paper instead of just throwing them away.

Recycling is more expensive than just burning garbage. But burning garbage has a hidden cost. That cost is air pollution. Burning garbage puts carbon dioxide and other gases in the air. So we have a choice — to pay now for recycling or to pay the hidden costs of bad air.

These are changes people have made to protect the environment and to stop the greenhouse effect from getting worse. There are other changes we will have to make if the world does actually grow warmer.

Global warming does not mean that we would die from heat. The average temperature might be 3 to 15 degrees higher. And this would not happen all at once. The changes would take at least 50 or 60 years.

What would warmer weather mean to you? You might have to wear different clothes. There would be less water for swimming

Recycling, rather than burning, paper is one way to reduce carbon dioxide in the air . . . and a way to save our forests.

pools. There would probably be no water for lawns. Maybe you would replace your lawn with a rock garden.

The food you eat would change. Farmers would not have enough water to irrigate crops. That might mean less fruit in the stores. You might eat tomatoes for only a few months of the year.

If you live in a cold climate, your winters might not be as cold. There might not be as much snow as there is now. That would make skiing or sledding or building snow forts difficult.

If you live in a warm climate, you might expect it to be warmer. You might want to use more air conditioning. But there may not be enough electricity to do that.

Do you like big cars? Big cars burn a lot of gasoline. They also send lots of carbon dioxide into the air. Once the greenhouse effect took hold, there would not be many big cars. Driving a big car might even be against the law. You might not even have a car. You might ride a bicycle or walk a lot more. And you would probably use public transportation to go on long trips.

The work world would change, too. With fewer forests, there would be fewer jobs for forest rangers. There might be fewer jobs for farmers in Iowa. But there might be more farming in Canada.

There might be new kinds of jobs. As we do more recycling, more jobs will be created in the recycling industry. Global warming would bring new industries. These industries would try to protect air and water. So there might be more jobs in conservation work.

By using public transportation, people can help decrease auto emissions and thereby slow down the greenhouse effect.

STOPPING THE GREENHOUSE EFFECT

What will happen if we make no changes at all in our lives and laws? Some scientists say that the temperature will rise at least 5 degrees in 50 years. Others think that it might rise 15 degrees. If we use more energy than we do today, the changes will be worse. Then we might see as much as a 30-degree rise in temperature.

On the other hand, we can try to slow the warming. Even if we try very hard, temperatures will still rise some. But they might rise only by 3 or 4 degrees. And, of course, all these changes will take place over a number of years, not overnight.

What kinds of changes will help to slow the greenhouse effect? What kinds of actions will speed it up? Putting more carbon dioxide (and other greenhouse gases) in the air will speed up global warming. Putting less carbon dioxide in the air will help to slow it down. And taking carbon dioxide out of the air will help even more.

We can ask these questions in another way. What kinds of actions put carbon dioxide in the air? What action could be taken to put less in the air? And what could be done to take the gas out of the air?

Driving cars is one of the chief ways of putting carbon dioxide in the air. Some experts say that cars are responsible for half the carbon dioxide we put in the air in the United States each year. Cars that get good gas mileage burn less gasoline. They release less carbon dioxide into the air. This means that one way to protect the air is to drive more efficient cars. Some engineers are at work on new cars. They are testing cars that would go one hundred miles

This restaurant has installed solar panels to capture the energy of the sun and turn it into electricity.

on one gallon of gasoline. Others are working on cars that run on batteries.

Another way to protect the air is to drive less. People can form car pools. Or they can walk and ride bicycles some of the time. In many other countries, people drive less and ride bicycles much more than we do. Mass transit is also part of a solution. Trains and subways can carry lots of people without using as much fuel.

Power plants are another source of carbon dioxide. Many power plants burn coal or oil to make electricity. The smoke and gas that these plants put into the air are called emissions. The emissions send lots of carbon dioxide into the air. Coal and oil are called "fossil fuels." One way to cut down on this carbon dioxide is to burn fewer fossil fuels. Another way is to use less electricity.

People are already using solar power to produce electricity. It does not produce any carbon dioxide. They are using waterpower as well. Wind power is another "clean" source of power. Geothermal power means power from the heat in the earth. Nicaraguans are already using it. They are making electric power from the heat of a volcano. Some places have lots of wind. Others have lots of sun or water. No single answer will be right for every place.

Nuclear power does not produce carbon dioxide, either. Some people say this is a good way to stop using fossil fuels. Nuclear power does have other problems. Its safety is often challenged. Many people think that changing to nuclear power would just be changing an old problem for a new one.

Using other fuels is not always easy. If a power plant is built to burn fossil fuels, changing to wind power is not likely to work. That power plant must keep on burning fossil fuels or shut down. But there are ways to cut down on carbon dioxide from the fuels. Some fossil fuels are cleaner than others. Cleaner fuels burn with-

Solar energy is one alternative to burning fossil fuels that damage the environment. This house runs on solar power.

out producing as much carbon dioxide. Some plants can switch to cleaner fuels.

If we use less electricity, then we will not need as many power plants. There are some easy ways to use less electricity. We can put on sweaters in the winter instead of turning up the heat. We can live with warmer rooms in the summer. These are small but important steps.

Some communities are turning garbage into a source of electricity. Columbus, Ohio, burns 2,000 tons of garbage a day in a waste-incineration plant. The plant, in turn, produces electricity for the city. Baltimore and Detroit also have waste-incineration plants.

The windmill, the ancient device for creating energy out of breezes, is making a comeback. In Hawaii, for instance, wind energy provides 3 percent of the state's electrical needs. In Arizona, windmills produce energy for water pumps that take underground water to crops and livestock.

On October 24, 1988, *Time* magazine reported how one company compensated for the greenhouse gases its factory might put into the atmosphere. Applied Energy Services was constructing a coal-burning plant in Uncasville, Connecticut. At the same time, acting on the recommendation of the World Resources Institute, an environmental research center in Washington, D.C., they donated $2 million to CARE. The money was to be used for buying seeds to plant crops in Guatemala. There, 40,000 local farmers would use the 52 million seeds to plant crops. These crops would produce enough oxygen to absorb almost the exact amount of carbon dioxide the Uncasville plant might produce. This idea might be a model for companies of the future.

Changing the way buildings work can help, too. Today many buildings have windows that do not open. They rely on air conditioning and heating. Why not construct buildings with windows that open and shut? Then we could use open windows to cool buildings at least a part of the time.

There are other changes in buildings that can save energy. Good insulation can help a building to stay warmer in winter and cooler in summer.

All of these are ways to cut down on the amount of carbon dioxide we put in the air.

Of course, it is also possible to move in the other direction. If we put more carbon dioxide in the air, we will make global warming worse. Driving bigger cars can make it worse. Using more electricity can make it worse.

Finally, we can try to take some carbon dioxide out of the air. That is more difficult to do. The best way is by planting more trees. But growing trees takes a long time, and right now more trees are being cut down than are being planted. Whole forests are being destroyed each year.

Planting new forests is called reforestation. But today deforestation is happening ten times faster than reforestation. To slow down global warming, we need to plant more trees.

Of course, no one action can stop the greenhouse effect. It is a big project. It will require many changes in the way we live.

These young, green trees were planted after a fire destroyed the forest. Reforestation is one important way to protect the atmosphere.

MAKING NEW LAWS

The greenhouse effect is a problem for the whole world. One person can't stop it. One town can't stop it. Not even one country can stop it. To stop the greenhouse effect, people all over the world need to work together.

New laws can help end global warming. Laws can order car makers to make cars that get more miles per gallon. Laws can support research on wind power or sun power. Laws can make power plants clean up their emissions.

Today, almost every country has some environmental laws. The people of the Netherlands are so concerned about the environment that some Dutch political parties have fallen completely out of favor over environmental issues. In May 1989, the government in Holland collapsed over a plan that would have discouraged the use of cars because of carbon dioxide emissions. They have now passed laws that will reduce their carbon dioxide emissions by 8 percent by 1994.

The Dutch have a special reason for being concerned about the greenhouse effect. According to some scientists, the greenhouse effect will cause the sea to rise two to six feet over the next century. This would be devastating to the Netherlands, because two-fifths of its land—which took the Dutch people centuries to reclaim from the sea—could sink back under water.

A rising sea would affect other countries as well. If the sea were to rise only three feet, 10 percent of the population of Bangladesh, east of India, would drown, and one-fifth of Egypt's fertile land would be lost.

Taxes can help, too. Many countries have high taxes on gasoline. When gasoline costs more, people use much less of it. Costa Rica has high taxes on cars. It also has a very good bus system. That means that many people do not need cars. They are able to take buses instead.

Many poor countries do not use much electric power now. They want and need to develop power plants and factories. If they do not, then they will not have jobs for their people. These countries look at the rich countries of the world and say: "You already have lots of cars. You have lots of factories. You use lots of electricity. Your people live well. It is not fair to tell us that we must stay poor."

Rich countries of the world do use more power. They produce much more carbon dioxide than the poor countries. For example, the United States is a very rich country. Only 5 percent of the people in the world live in the United States. But the United States produces 25 percent of the carbon dioxide in the world. We put more than a billion tons of carbon dioxide into the air each year.

What is a fair way to solve this problem? Should we say that every country should cut down on the number of cars it has? Then the rich countries would still have lots of cars and the poor countries would not have many at all. Should we say that there can be no new coal-burning factories or power plants? Again, the rich countries already have these plants. The poor countries do not have nearly as many. The rich countries would have to give up a little. The poor countries would stay poor forever.

Fairness is a big problem. But there are other problems with making laws. Many people in government are more ready to talk than to act. Environmentalists say that neither the president nor Congress is doing much. Why not?

Politicians fear that laws like a higher gas tax would be unpopular with voters. Car makers fight against laws for cars with better mileage. Power plants and coal mines do not want laws for cleaner power plants. Many people don't want to spend their own

One easy way everyone can help protect the environment is by sorting trash. These glass bottles are ready to be recycled and used again.

money to clean up the air. Everyone would like someone else to do it. So not much happens.

Besides that, some people still say that we do not know enough about global warming. They say we need to study the greenhouse effect more. After we know more about it, then we can make laws.

One environmental leader says that is a bad argument. Tina Hobson is the director of Renew America. This group works for more energy efficiency. Ms. Hobson says: "If your child's temperature is 101 degrees, you do something about it. You don't wait until it reaches 107. By then it's too late."

WHAT YOU CAN DO

How can you help stop global warming? There are steps that every person can take. Both your actions and ideas are needed.

Cars are a big source of carbon dioxide in this country. You can think about ways to use cars less. Can you walk or ride your bicycle to the store? To a friend's house? To a softball game? If you can, that will save your parents a trip in the car. And it will save the air from some carbon dioxide. You can talk to your friends about walking or biking together. That will help even more to keep the air clean.

Car pools are another good way to use less gas. Sometimes parents can take four or five students to a game or to the library. That means one trip and one car instead of four or five. Sometimes adults can arrange to ride together in one car to get to and from work, too.

Saving energy helps to stop global warming. There are many small ways to save energy. Turning off lights when you leave a room is one way. Turning off the television or radio when no one is watching or listening is another way. Saving hot water saves energy, too. Did you know that taking a shower uses less water than taking a bath?

Recycling is another important way to save energy. When paper is recycled, that means fewer trees are cut down to make new paper. You can help by sorting paper, cans, and glass for recycling. This might even be a good project for your school. Your class could learn about recycling in your town. Then you can tell others. You can help to find ways for your school to recycle, too.

Since trees take carbon dioxide out of the air, planting trees is also helpful. Perhaps your school could start a tree-planting project. You might suggest this to your 4-H club or Girl Scouts or Boy Scouts or church.

You can't stop global warming alone. No one can. It will take all of us working together. Your work is an important part of the solution!

FOR MORE INFORMATION

Environmental Defense Fund
257 Park Avenue South
New York, NY 10010

Global ReLeaf
P.O. Box 2000
Washington, DC 20013

The Greenhouse Crisis Foundation
1130 17th Street NW
Suite 630
Washington, DC 20036

League of Conservation Voters
2000 L Street — Suite 804
Washington, DC 20036

The Union of Concerned Scientists
26 Church Street
Cambridge, MA 02238

FOR FURTHER READING

Dolan, Edward F. *Drought: The Past, Present, and Future Enemy.* New York: Franklin Watts, 1990.

Gay, Kathlyn. *The Greenhouse Effect.* New York: Franklin Watts, 1986.

———. *Ozone.* New York: Franklin Watts, 1989.

Gribbin, John R. *Future Weather and the Greenhouse Effect.* New York: Delacorte Press, 1982.

Koral, April. *Our Global Greenhouse.* New York: Franklin Watts, 1989.

Luoma, John R. *Troubled Skies, Troubled Waters.* New York: The Viking Press, 1984.

Sandak, Cass R. *A Reference Guide to Clean Air.* Hillside, NJ: Enslow Publishers, 1990.

GLOSSARY

atmosphere *The body of gases surrounding the earth.*
carbon dioxide *An invisible, odorless gas in the atmosphere which is a major cause of global warming.*
Celsius *A way of measuring temperature used by scientists and by many countries.*
chlorofluorocarbons (CFCs) *Human-made chemicals used in air conditioners, refrigerators, freezers, plastic foam, solvents, and some spray cans.*
deforestation *Destroying forests on a large scale.*
emissions *Gases given off by burning fuel, for example, in power plants or car engines.*
environmentalist *Someone who is concerned about clean air, clean water, and natural resources.*
Fahrenheit *A way of measuring temperature used in the United States.*
fermentation *Gradual decomposing of organic matter. Carbon dioxide is released during this process.*
fossil fuels *Fuels like coal, oil, or gas that have been formed underground for hundreds of thousands of years.*
geothermal *Taking heat from the earth.*
glacier *A large body of ice.*
global warming *Another name for the greenhouse effect, the theory that the earth is getting warmer because of human activity that has changed the atmosphere.*
greenhouse *A building made mostly of glass, used to grow plants year-round.*

greenhouse effect *The theory that the earth is getting warmer because of human activity that has changed the atmosphere.*

greenhouse gas *A gas that traps the sun's heat and holds it near the earth; especially carbon dioxide, nitrous oxide, methane, and CFCs.*

infrared *A kind of light that has a wavelength longer than that of visible red light and radiates heat.*

lava *Fluid rock that comes from a volcano and then hardens.*

methane *A colorless, odorless, flammable gas.*

nitrous oxide *A gas often used to put someone to sleep, also called "laughing gas."*

ozone *A kind of oxygen molecule.*

photosynthesis *The process by which plants use carbon dioxide and sunlight to form sugars and emit oxygen.*

reforestation *Systematic replanting of forests and trees.*

solar power *Power from sunlight.*

theory *An explanation of how facts fit together and a guess at what can be expected to happen.*

ultraviolet *A kind of light that has a wavelength shorter than visible light and contributes to sunburn.*

INDEX

Arizona 34
Arrhenius, Svante August 16
atmosphere 6, 8, 11, 12, 14, 16, 19, 21, 34, 36

Bangladesh 37

Canada 23, 29
carbon dioxide 6, 8, 11, 12, 14, 16, 19, 26, 29, 30, 32, 34, 35, 37, 38, 41, 42
Celsius 18, 21
chlorofluorocarbons (CFCs) 11, 12, 14
Connecticut 34
Costa Rica 37

deforestation 14, 35

Egypt 24, 37
emissions 29, 32, 37
environmentalist 21, 23, 38

Fahrenheit 18
fermentation 6
flat earth theory 8

fossil fuels 32

geothermal power 32
glacier 16, 17
global warming 6, 16, 19, 21, 23, 24, 26, 29, 30, 35, 37, 41, 42
Goddard Institute 16, 18
greenhouse 5, 6, 8, 11
greenhouse gas 11, 12, 14, 18, 19, 30, 34
Guatemala 34

Hansen, James 21
Hawaii 16, 34
Hobson, Tina 41

India 37
infrared rays 11
Iowa 23, 29

Kansas 23

laughing gas 11
lava 16
Louisiana 23

methane 11
Minnesota 23

Netherlands, the 37
New York 23
Nile Valley 24
nitrous oxide 11
North Pole 18, 23
nuclear power 32

Ohio 34
oxygen 8, 12, 14
ozone 12

photosynthesis 8

recycling 26, 29, 42
reforestation 35, 36

Sahara Desert 24
Siberia 23
solar power 32, 37
Solow, Andrew 21
South Pole 18, 23

theory 8, 16

ultraviolet rays 11, 12

Washington, D.C. 19, 34
wind power 32, 34, 37
Woods Hole Oceanographic
 Institution 21